FAÇADE-RESTAURANT

餐厅店面

《新空间》编辑组 编
张晨 译

辽宁科学技术出版社

FAÇADE-RESTAURANT
餐厅店面

Bz SE ARRIENDA
627 53 75 23
ORFILA
PaCatar

005

ORFILA

30 ml
caffè

009

30
ml
caffè

30
caffè

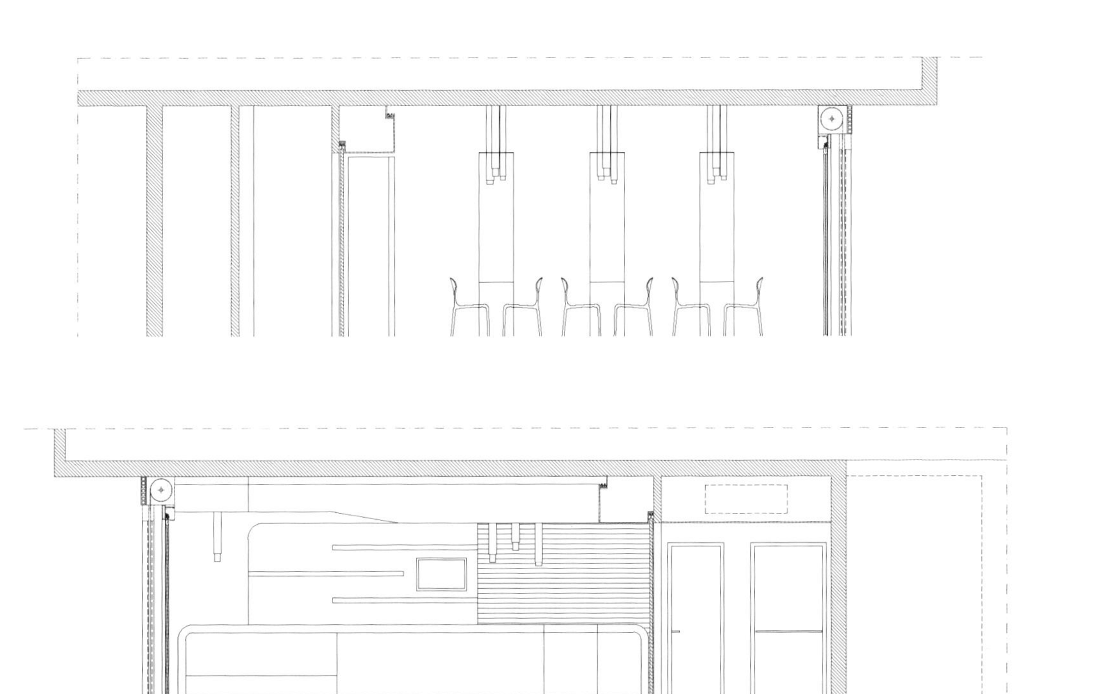

78
5 6 0
restaurante

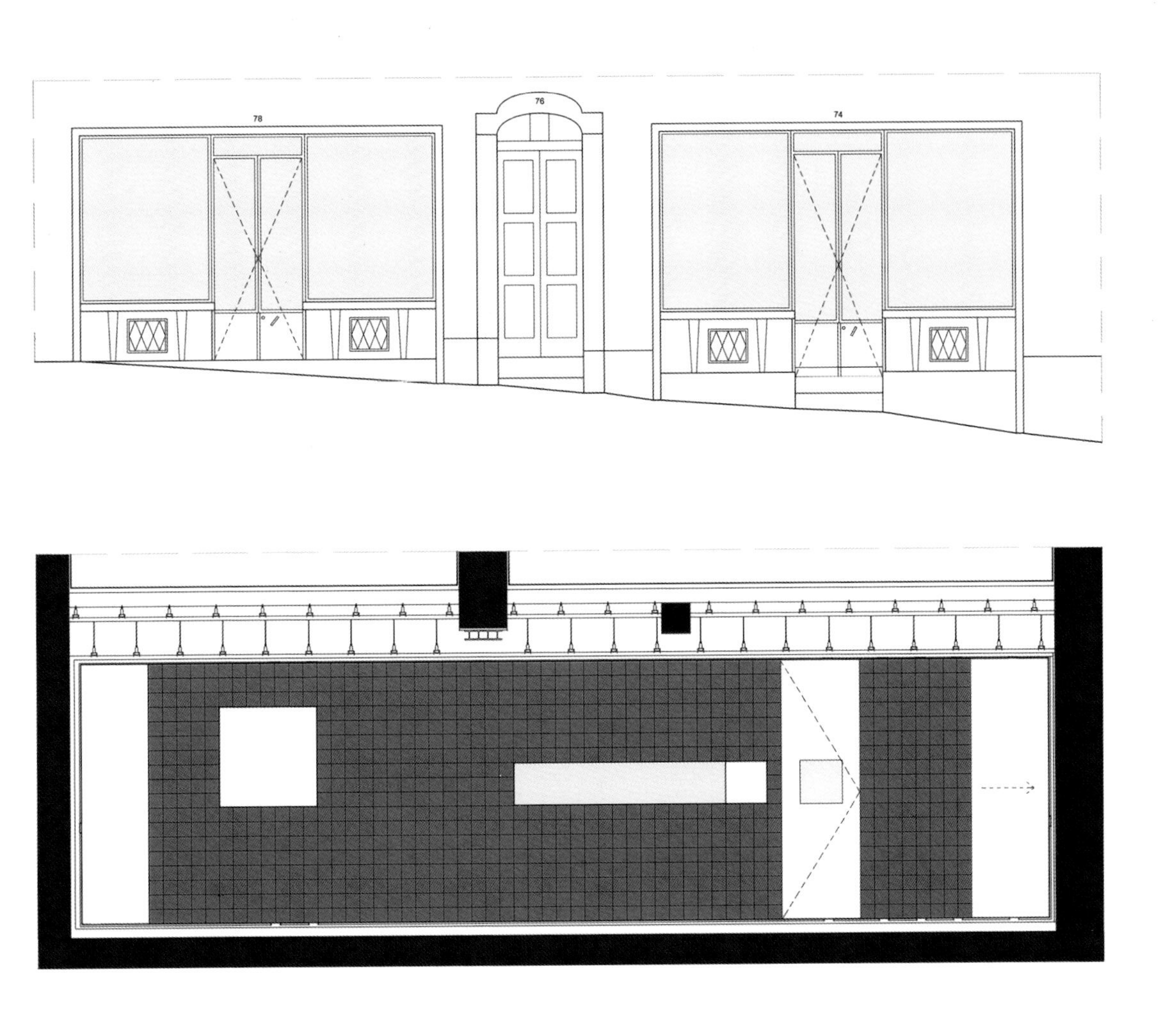
78
76
74

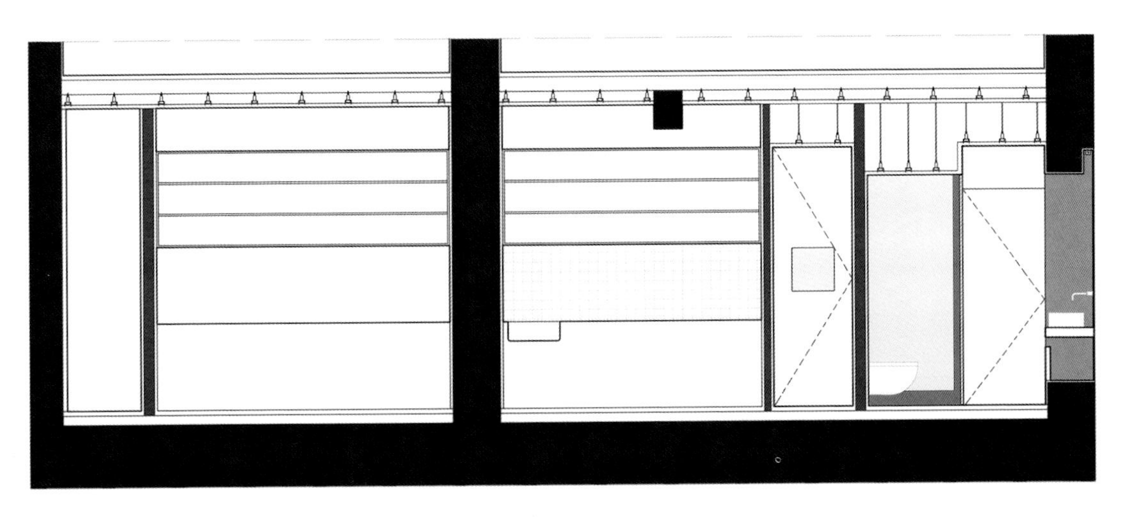

5 6 0
restaurante

αϊνικόλας
αϊνικόλας

ANNAM VIETNAMESE CUISINE

AN
NAM

AN
NAM
VIETNAMESE CUISINE

café
bubbletease
11

COCOS EINFACH

COCOS EINFACH ASIATISCH

COCOS
EINFACH ASIATISCH

FITCH & SHUI
ENTREPRENEURS

soundCloud
facebook
androidgames

BASE
CAMP
COFFEEBAR
by mokador

BASE
BASECAMP
BASE
CAMP

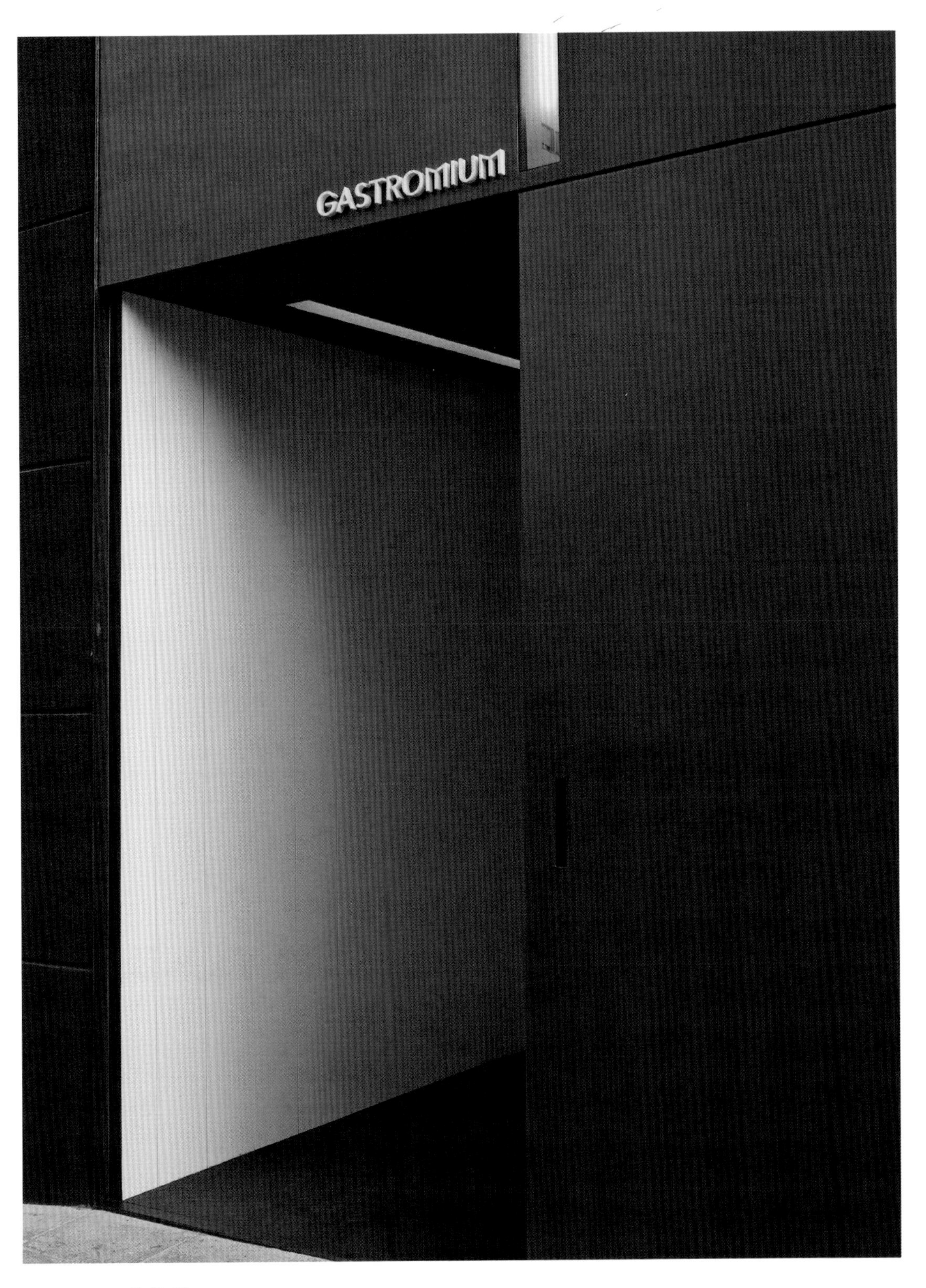
GASTROMIUM

BEMBOS

BEMBOS
park
CIRCUS

BEMBOS

BORN
THIS WAY
TONIGHT
STAR
STAR

FROZEN

BEMBOS

BEMBOS
toma lo bueno

BEMBOS

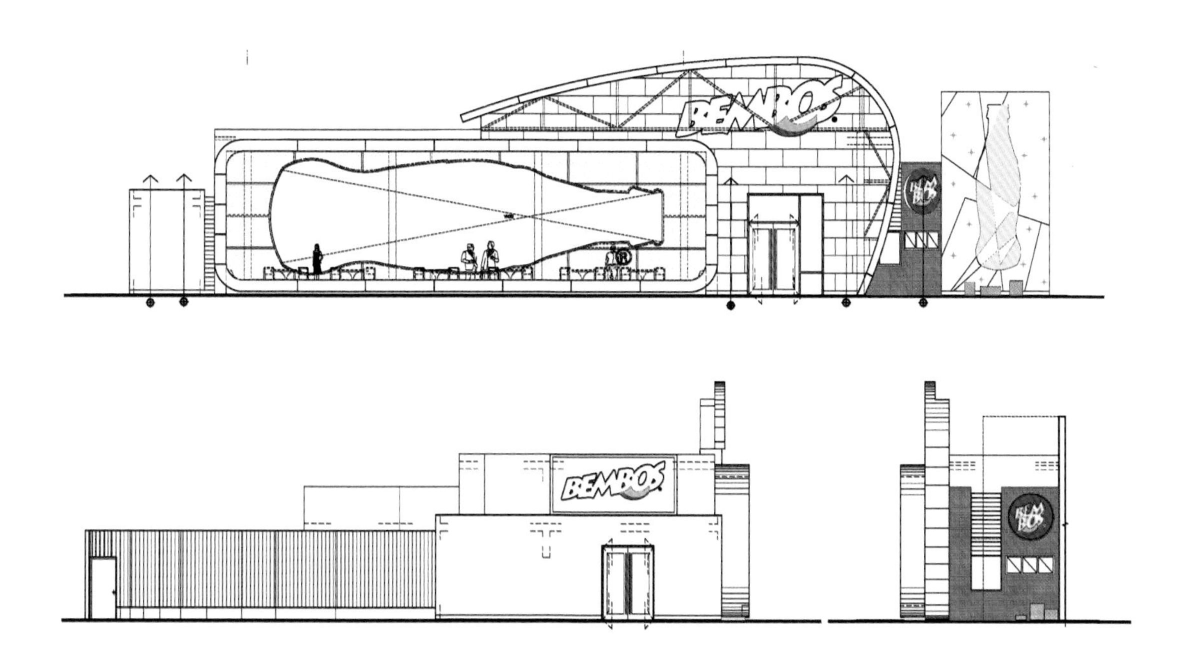
BEMBOS
BEMBOS

¡BEMBONA!

Cielo
eleva tu vida
Cielo
eleva tu vida
the
Search

SABORES
DE MI PERÚ
desde
S/.4.90
regular
BEMBOS
BUTIFARRA
Cocona
y cecina
BEMBOS

CAMINOS DEL INCA
AD
Actually designs
ESTACIONAMIENTO GRATUITO
POR COMPRAS EN EL 1° Y 2° NIVEL

SABORES
DE MI PERÚ
desde
S/.4.90
regular
BEMBOS
BUTIFARRA
Cocona
y cecina
BEMBOS
BEM
BOS

SABORES
DE MI PERÚ
Cocona y cocina
BEMBOS

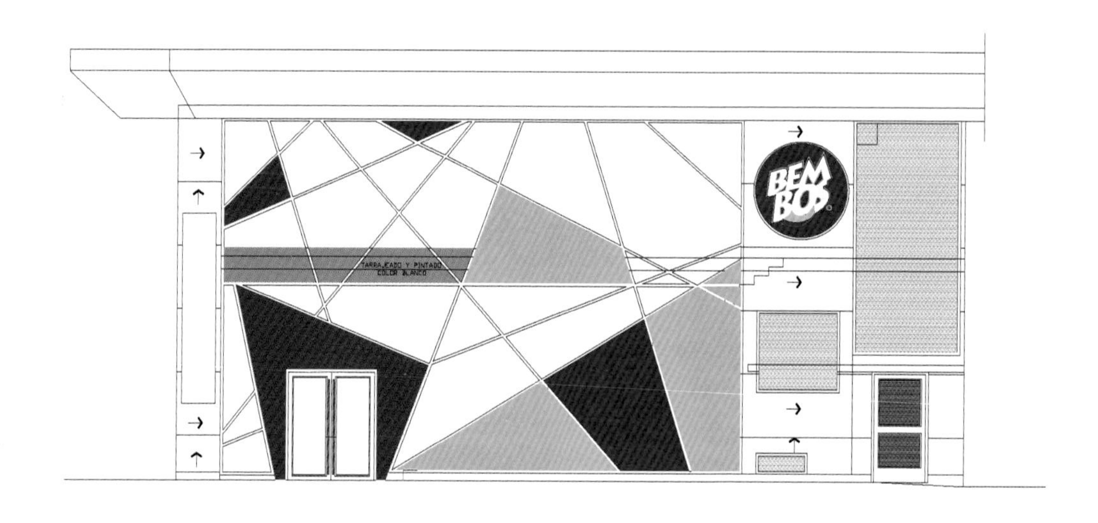
BEM BOS
TARRAJEADO Y PINTADO
COLOR BLANCO

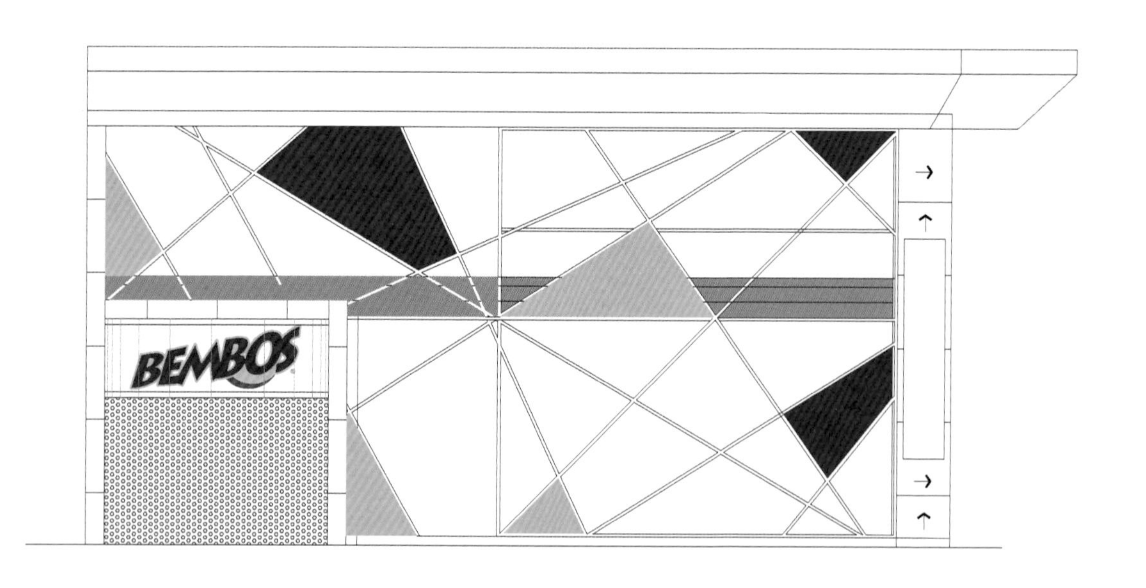
BEMBOS

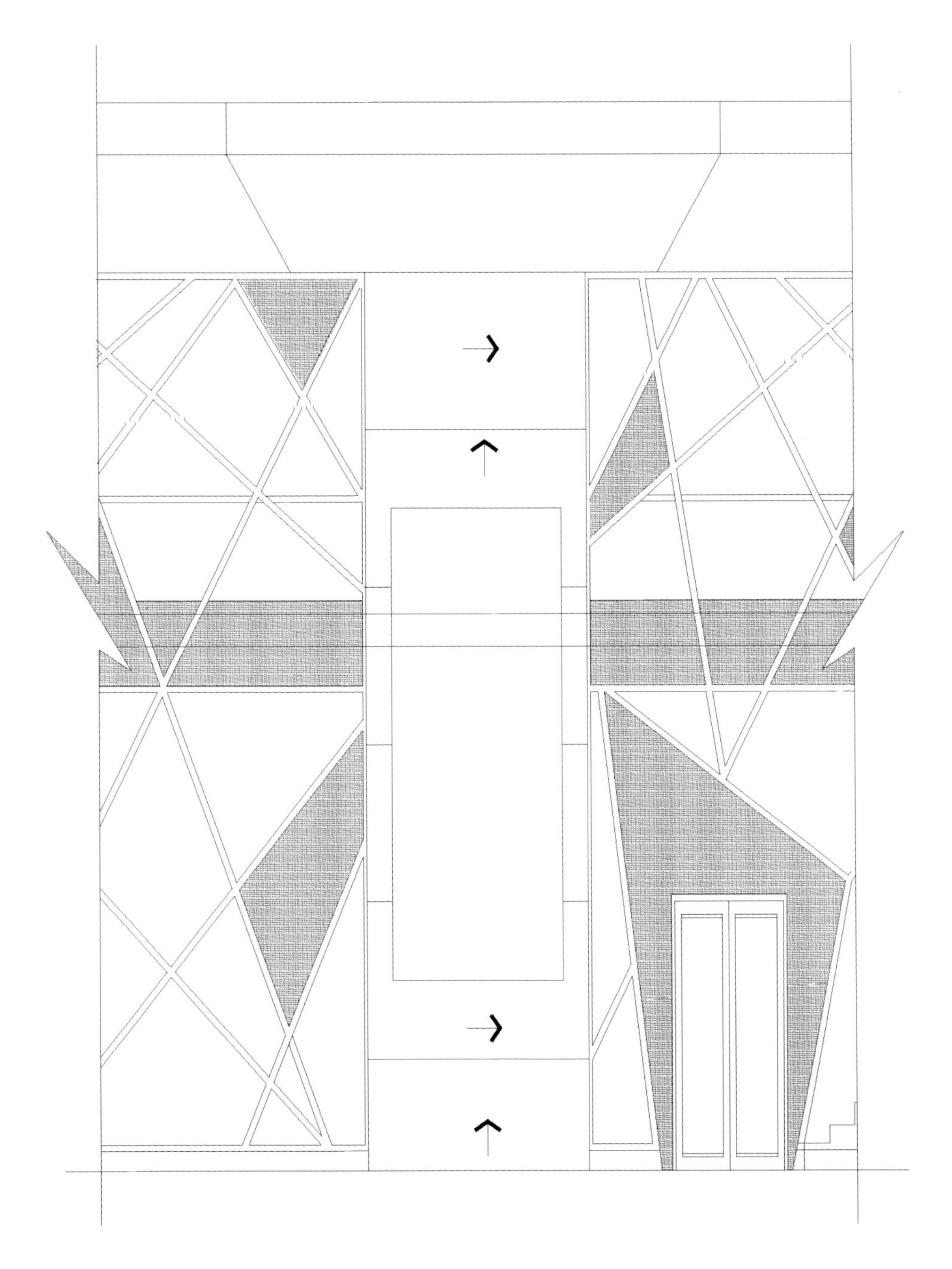

EXPRESATE

BEMBOS
AUTO BEMBOS
BEMBOS
park

INGRESO VEHICULAR

BEMBOS
BEMBOS
park

BEMBOS

BEMBOS
AUTO-BEMBOS
BEMBOS
INGRESO VEHICULAR

MIRAFLORES
Desayunos

Desayunos

SALIDA

Banco de Crédito BCP
VEHICULAR

BURGER KING

BURGER
KING

EXIT
BURGER
KING

Have it You
BREAKFEAST PLATTER
THE MUSHROOM SWISS FROM BURGER KING
WITH THE KING, YOU CAN.
221184
WHOPPERBILITIES
100% FLAME-BROILED BEEF

bok

fach asia

CAFE 501

CAFE
DEL
ARCO

CAFE
DEL
ARCO

CAFE
DEL
ARCO

Lindt

Lindt
LINDT & SPRÜNGLI
MAÎTRE CHOCOLATIER SUISSE DEPUIS 1845
Lindt

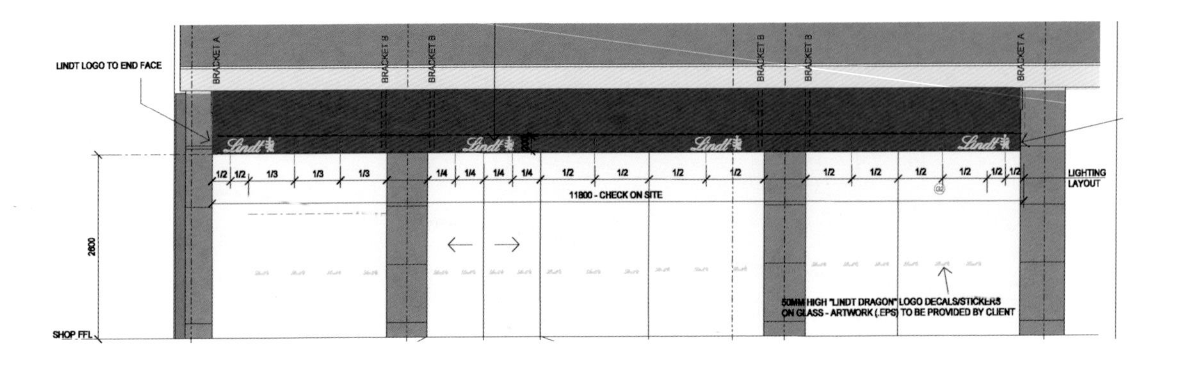

01 AWNING - ELEVATION 1:50 ON A3

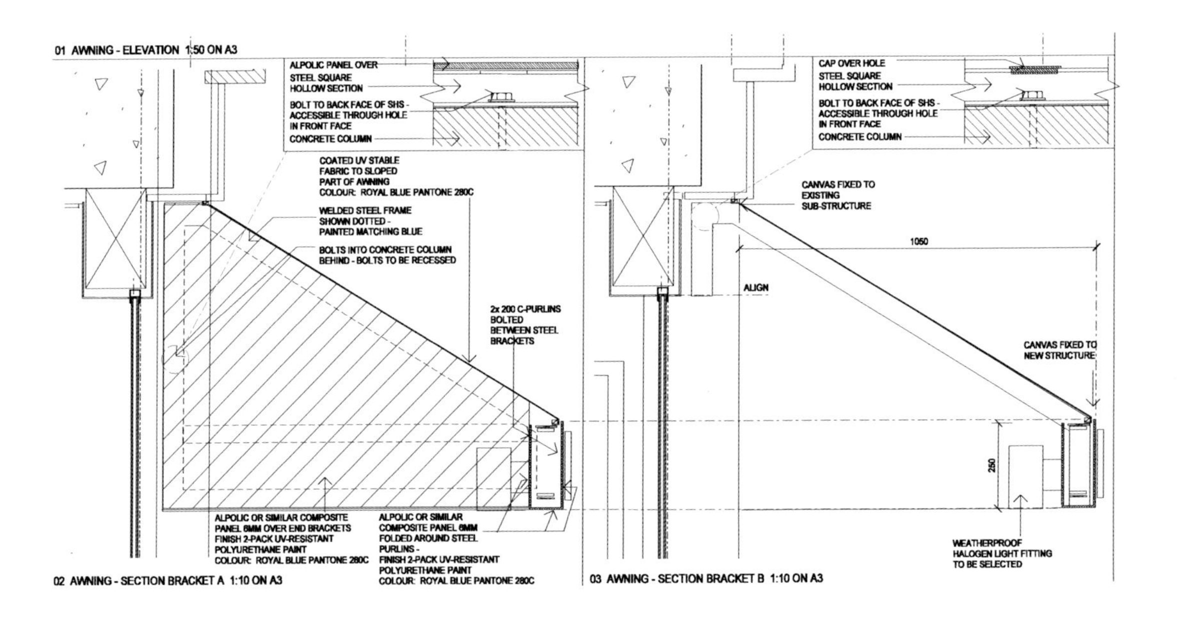

02 AWNING - SECTION BRACKET A 1:10 ON A3

03 AWNING - SECTION BRACKET B 1:10 ON A3

Lindt
Lindt

香記
ASIAN SWEETS AND CAFE KOKI
香記
ASIAN SWEETS AND CAFE KOKI

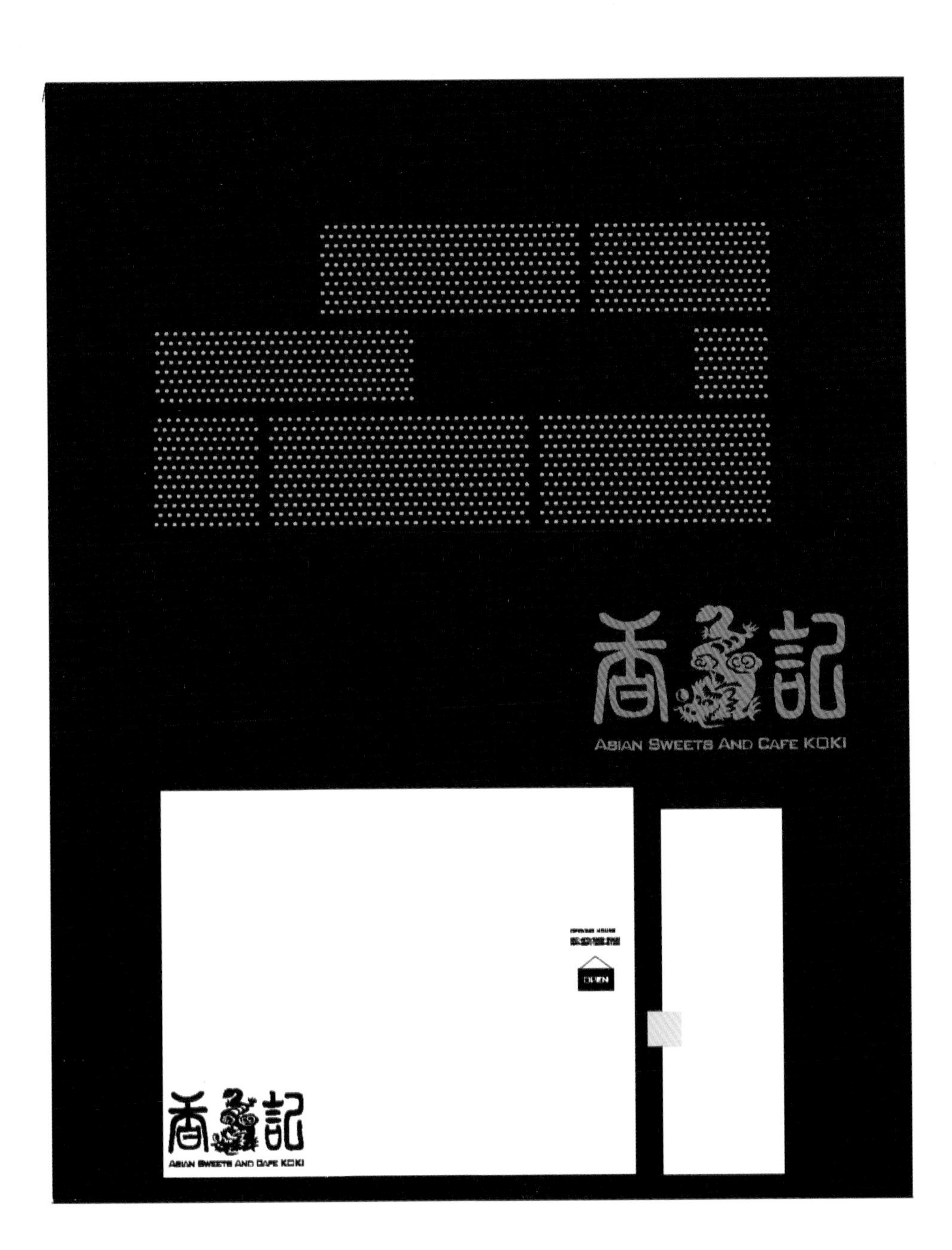

香記
ASIAN SWEETS AND CAFE KOKI
OPEN
香記
ASIAN SWEETS AND CAFE KOKI

Café Ritrovo
Café Ritro

Special
Big Breakfast $9.90
(Bacon, eggs, tomato
hash brown, toast)
Chicken Schnitzed
with chips and salad
$12.90
nourish
VICTORIA ARDUINO
Chicken mushroom
Beef mushroom
Bacon and cheese
Ritrovo's Pizzas
BBQ Chicken
Hawaiian $7.90
Vegetarian

CHEF'S TABL

56

co
co
ro

cocoro
new style
cuisine

FRANKFURT
STATION
FST
HOT DOGS
BURGERS
OPEN
MON-SUN
10.00-03.00

DELICATESSEN

Wallpaper*
FRAME
frieze
Wallpaper*

dimsum
17

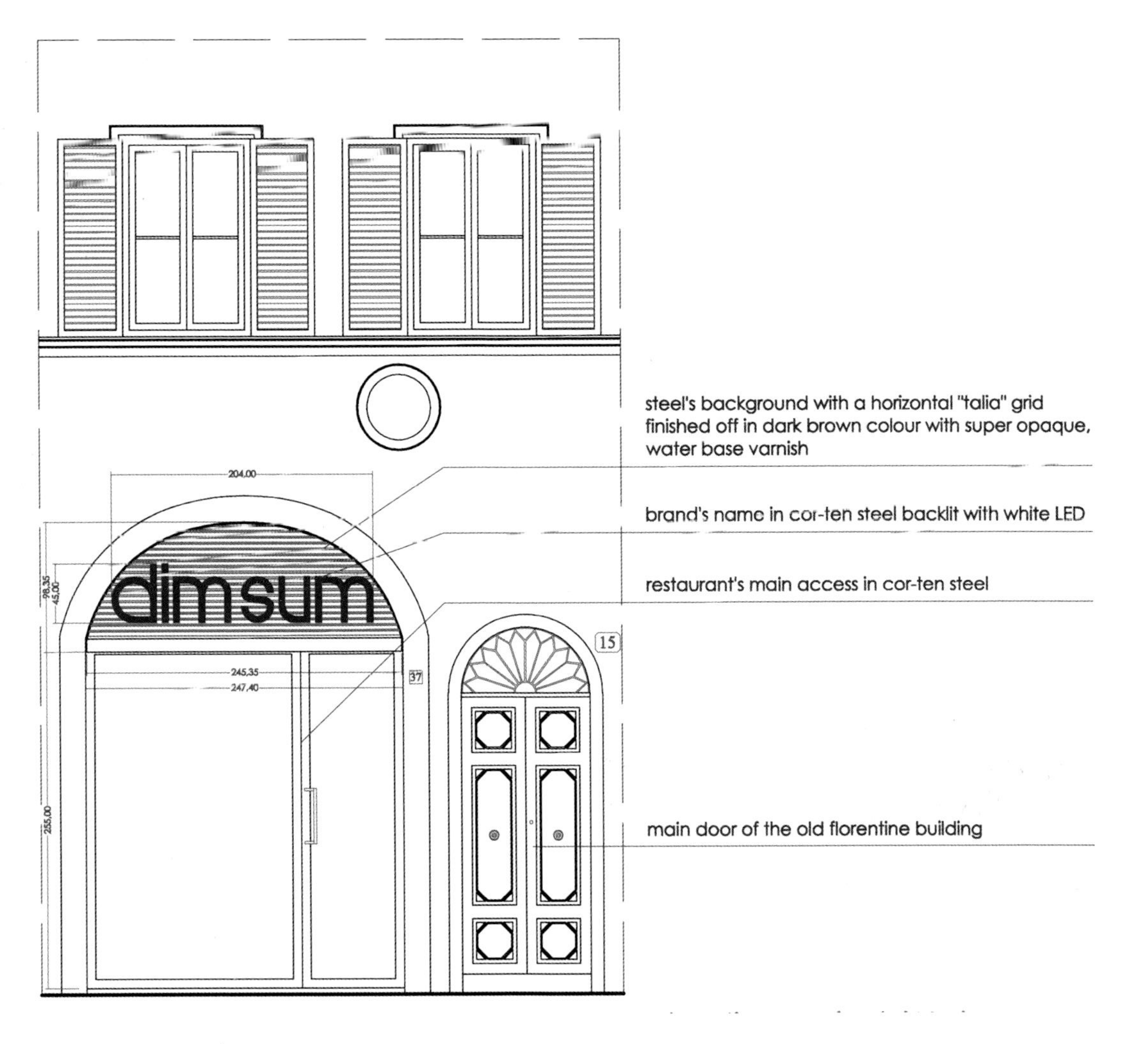
steel's background with a horizontal "talia" grid finished off in dark brown colour with super opaque, water base varnish
brand's name in cor-ten steel backlit with white LED
restaurant's main access in cor-ten steel
main door of the old florentine building
dimsum
204,00
98,35
45,00
245,35
247,40
255,00
37
15

ENTRANCE
BEWARE OF THE TRAINS
PULL
DINING CAR
I
G.I.P.
01975

GREAT INDIAN PENINSULA
DINING CAR
Nº 12
DINING CAR
THE EXPLORERS CLUB
FIRST
I
FIRST
FIRST

Nº 12
THE EXPLORERS CLUB

INDIAN PENINSULA
DINING CAR
Nº 12
THE EXPLORERS CLUB
FIRST
II
IV
THE EXPLORERS CLUB

φάρμα

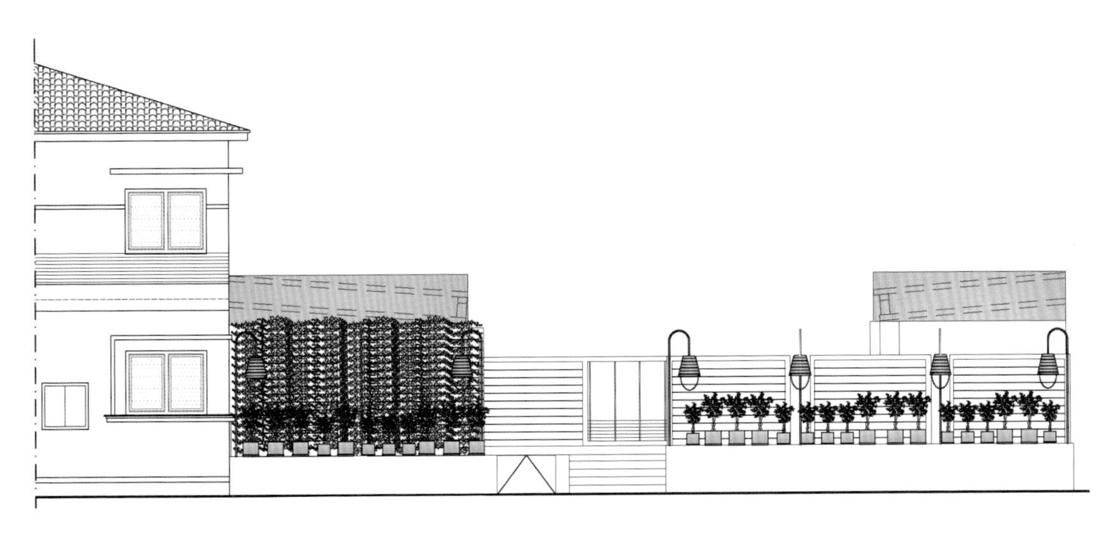

φάρμα

Franzesco

DEICHMANN
DEICHMANN
back WERK
back WERK
back WERK

ZZA
ricciosa
fiosi

fu
סושי

Fuchsstube

fuchsbau

FUGA

o eat

Shots
FUGA

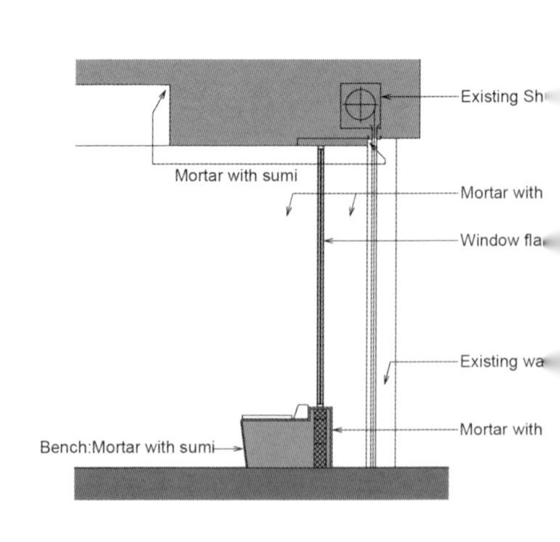
Existing Sh
Mortar with sumi
Mortar with
Window fla
Existing wa
Mortar with
Bench:Mortar with sumi

GEEK
COMFORTABLE
BAR & CAFE

銀座
ステーキ
TAJIMA

GIORNO
TRATTORIA/BAR
GIORNO
TRATTORIA/BAR

GIORNO
TRATTORIA/BAR

GIORNO
TRATTORIA/BAR

GIORNO
TRATTORIA/BAR

Giuseppe
STUDIO
ENTERTAINMENT
STORE

aldo & Sons

Giuseppe Arnaldo & Sons

goodcals
PUSH
Green
& Marine

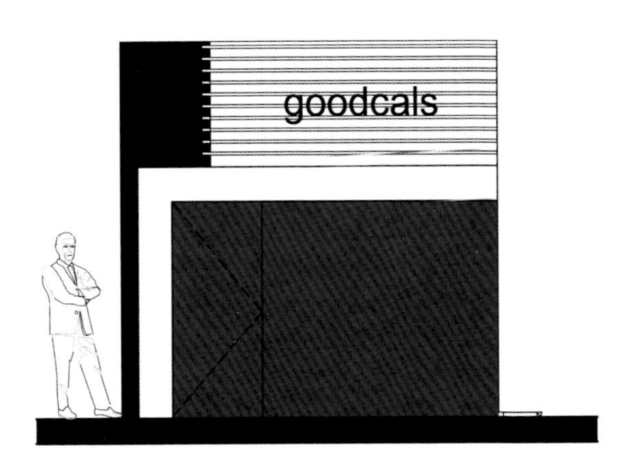
goodcals

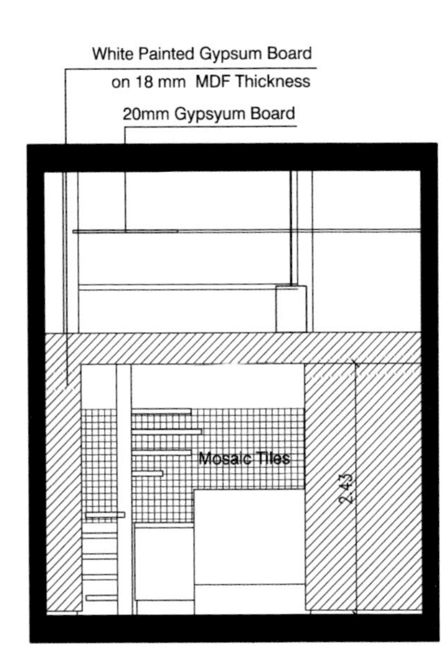
White Painted Gypsum Board
on 18 mm MDF Thickness
20mm Gypsyum Board
Mosaic Tiles
2.43

goodeals

+ and green
+g
and green

Restaurant & Bar
134-136

inamo

inamo
Restaurant & Bar
134-136

Restaurant & Bar
134 - 136

ANESE BISTRO and TAPAS
JAPANESE

STRO and TAPAS
O and TAPAS
JAPANESE BISTRO and TAPAS

LEANERS
LORS 752-8253
CLEANERS
CLEANERS
Naya
EXIT

WE SELL
WATCH BATTERIES
&
WATCH BANDS

olivino
olivo

10
olivomare
8
A

olivino
12
olivino

Olivocarne
Olivocarne
BANHAM

Olivocarne
Olivocarne
DE VROOMEN
BANHAM
61

ARATA
新
ARATA

barbacco
ENO TRATTORIA
LUNCH
DINNER
TAKE-OUT
CATERING

おかゆと麺の店
かゆさんちん
粥餐庁

おかゆと麺の店
粥餐庁

Trattoria
Kodama

30
OMMERY
La Bonne Bouche

LA BONNE BOUCHE
La Bonne Bouche
BISTRO
PLAT DU JOUR
BAR
Grolsch
GRANDE SÉLECTION

La Galette

La Galette

La Galette

CALLE
DE
SORNI
La Petite Brioche
Bakery
Horario
Martes
Domingo
Prueba
nuestros
Croissants
rellenos de
Frambuesa

La Petite
Brioche
Bakery
9 → 20h
9 → 14h

CALLE DE SORNI
La Petite Brioche
Bakery
La Petite Brioche
Bakery
Martes a sabado 9→20h
Domingo y Lunes 9→14h
Prueba
nuestros
Croissants
rellenos de
Frambuesa

LaOliva

1F LaOliva
BF
LaOli

la Oliva

BRUNO
CASA DELLA PIZZA
BRUNO
CASA DELLA PIZZA

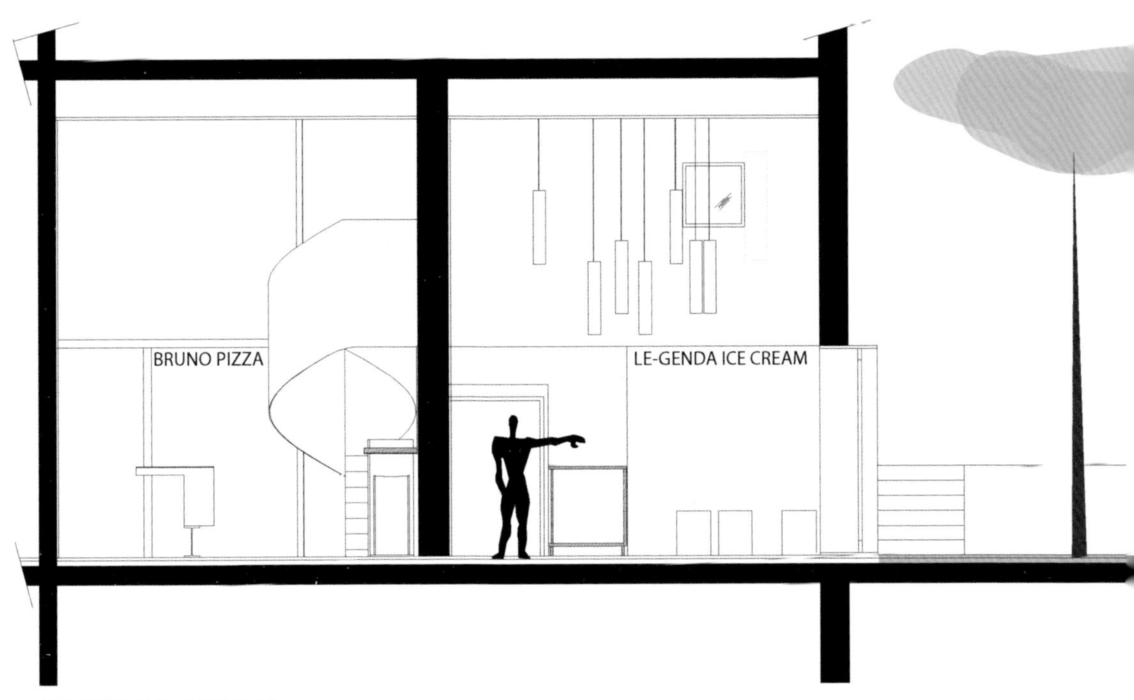

CROSS SECTION - SCALE 1:25

Les Bébés
CAFE & BAR

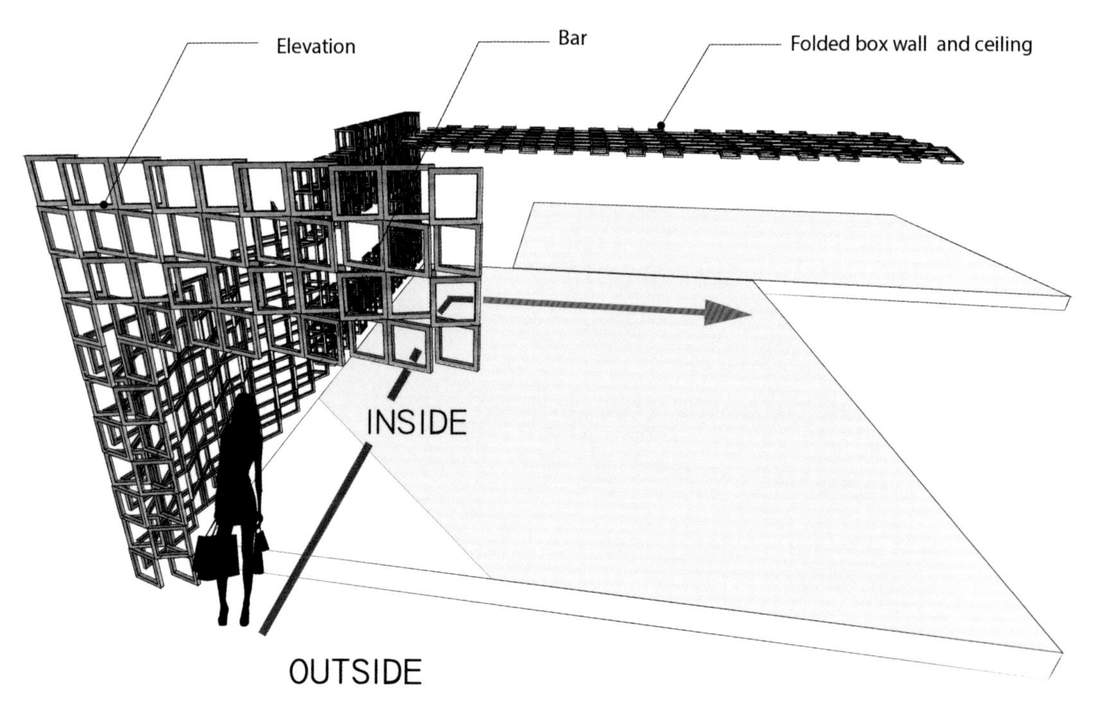
Elevation
Bar
Folded box wall and ceiling
INSIDE
OUTSIDE

LITTL

OPERATION

CHEF

Coca-Cola

TWENTY
FIVE
LUSK
25

PACKING & PROVISION CO

mangiare
[Italian for eat]
mang
SECOM

for eat]
mangiare
mangiare
mangiare
mangiare
mangiare
mangiare
PASTA
PIZZA

PASTA
from £4.50
PIZZA
[2 slices]
from £3.25
Ask for
Today's Special

all
made

master food restaurant | juice bar
master food restaurant
Pebble Peach $28
Berry Creek $28
Mango Mission $25
Pineapple Express $35
All Fruit $35
Fit 'n Fruitful $35

114
MAZZO

SAPORI
Sanbitter Sanbitter

MAZZO

McDonald's
easy morning
UNSER ANGEBOT

VI., JAKOMINI
Jakominiplatz
VI., JAKOMINI
Schönaugasse
1
ausgenommen

McDonald's
Jakominiplatz
easy morning
UNSER ANGEBOT

McDonald's
Jakominiplatz
easy morning
UNSER ANGEBOT

McDonald's
McCafé
McDonald's

McDonald's
McDonald's
McCafé

McCafé

cDonald's

DRIVE
DRIVE
McDonald's
McCafé
McCafé
McDonald's
McDonald's

More Café

More Café

Giolitti
Roma dal 1900
Giolitti

Giolitti
Giolitti
Roma dal 1900

Giolitti
Roma dal 1900
Giolitti Roma dal 1900
© Temmuz Arsiray

Giolitti
Roma dal 1900
Giolitti Roma dal 1900
© Temmuz Arsiray

nana's green tea

nana's green tea

nana's green tea

nana's green tea

nana's green tea
nana's green tea

nana's green tea
nana's green tea

Now Grilling

Now Grilling
Nando's
rcelos Cockere
his tale dates back to the 14th Century and lik
egends the details differ depending on who's do
the telling. Here's our version. A pilgrim was
passing through the village of Barcelos in Portugal
when he was wrongly accused of theft. This was a se
charge for which a guilty verdict meant certain death
The pilgrim was brought before the town's judge
who was about to eat a roast cockerel for dinner.
Feeling vulnerable in a strange village and knowing what
his sorry fate might be, the pilgrim pleaded:
If I am innocent, may that
cockerel get up and crow!
No sooner had he spoken, than the cockerel
got up and crowed heartily (well it is a legend!)
with that, the pilgrim was pardoned and allowed to go
on his way. Ever since, the Barcelos Cockerel has been
a symbol of faith, justice and good fortune.

NOPI

PANINO GIUSTO
Milano 1979
PANINO GIUSTO
Milano 1979

PESCADOS CAPITALES

SOBER
IRA BIA
GULA AVA
LUJURIA
RICIA
Apetito desmedido de comer y beber.
Apetito exacerbado de los deleites carnales.
Pasión del alma que causa indignación o enojo

SOBER
IRA BIA
GULA AVA
LUJURIA
RICIA

pulcinella
PIADINERIA
pulcinella
PIADINERIA

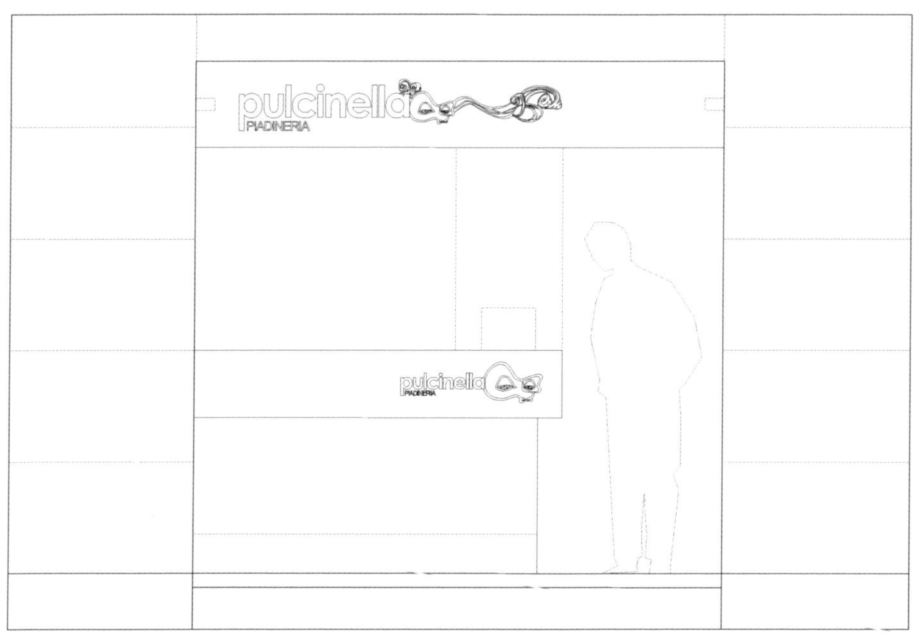
pulcinella
PIADINERIA
pulcinella
PIADINERIA

pulcinella
PIADINERIA

ORIGINAL TO THE CRUST
Pizza Hut
RESTAURANT
stuffed crust
Go Freestyle
pizzas & pastas
SALAD
dressed to impress
So come in
Pizza Hut
EAT
Drink

Pizza Hut
RESTAURANT

PizzaExpress, Richmond.
We're open
all day.
All day.
Every day.
From
8.45am.
PIZZAEXPRESS
PIZZAEXPRESS
PIZZAEXPRESS

LA TAVERNA DI POLDO

pizzazaza

pizzazaza

pizzazaza

youngeyes
POLAR
POL
Board Gam
POLAR
Board Games & Cafe
Tel: 083-138

R
afe'
OLAR

The
Beautiful Beer Garden
Great Function Room
The Princess
The Princess
The Princess
No smoking or drinking on fitzroy road please
Great British Food
Quality British Produce

FITZROY
ROAD NW1
cess of Wales
Fine Wines
Refreshing Beers
Proper Coffee
The Princess
The Princess
The Princess
The Princess
Princess of Wales
PLEASE HAVE RESPECT FOR OUR NEIGHBOURS WHEN LEAVING THESE PREMISES
THANK YOU

P.S.
PostScript
又及餐厅

Cielito QUERIDO CAFÉ
Cielito
QUERIDO
CAFÉ

Cielito
QUERIDO CAFE
café CON LECHE
TÉS
97¢
DE AQUÍ SOY
TÉS
LECHE
AZÚCAR
galletas

KITCHEN

R+D R+D
1323

R+D
KITCHEN
R+D
KITCHEN
SANTA MONICA
NEWPORT BEACH
PARK CITIES
1323

POW WOW
COFFEE
ESPRESSO
ESPRESSO BASED DRINKS
afri-cola
afri-cola

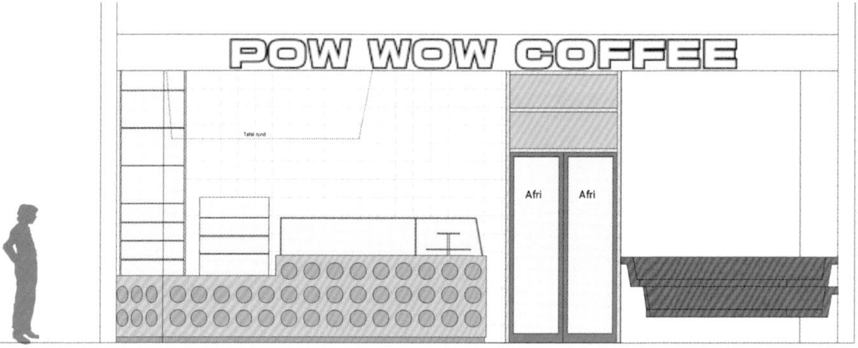
POW WOW COFFEE
Tafel rund
Afri
Afri

SILK ROAD

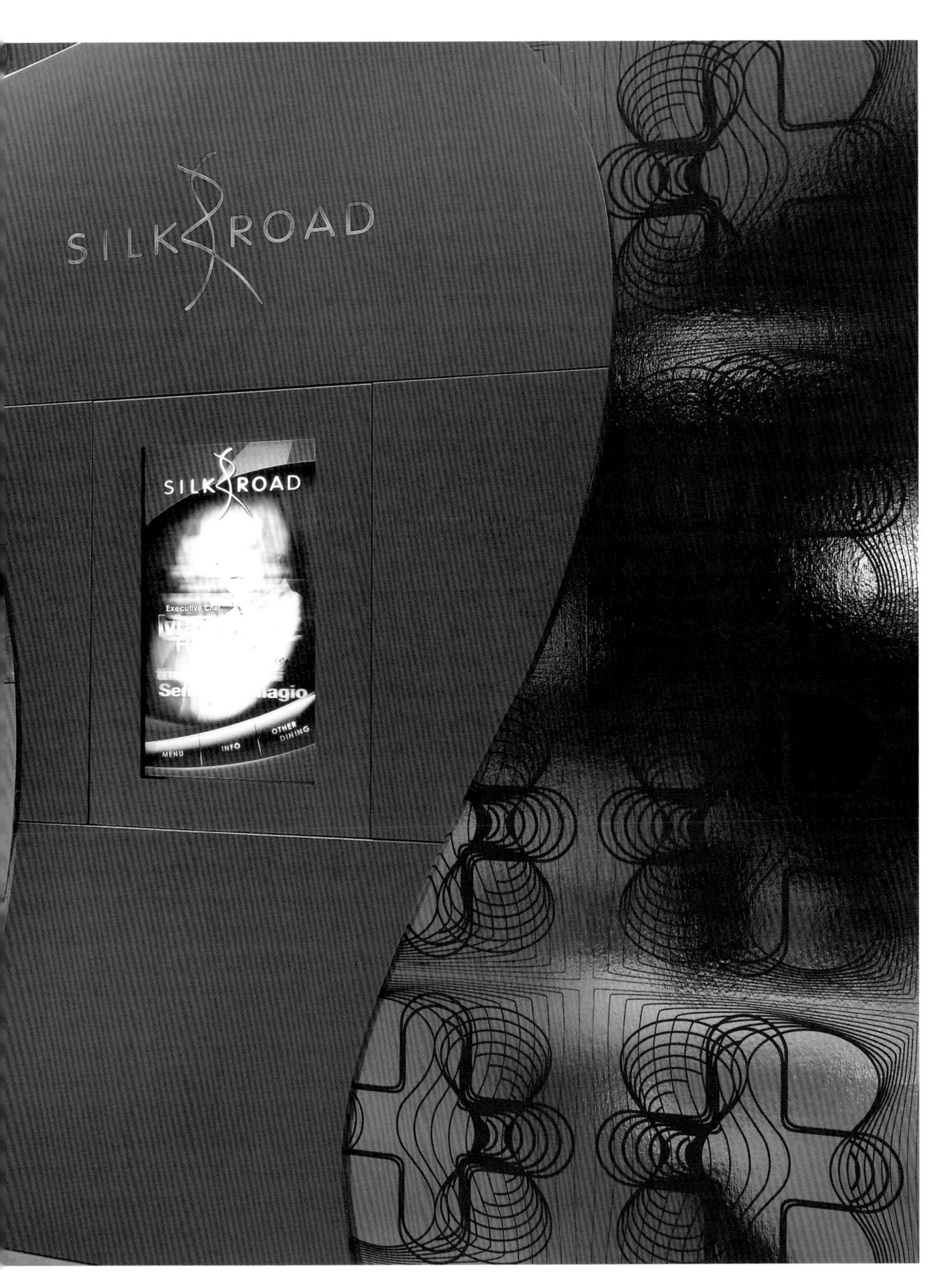
SILK ROAD
SILK ROAD
Executive Chef
MENU
INFO
OTHER DINING

sāwāsdeLight

sm
SantaMarta

330
SOUPS

SPORTS
SOUP'S
GRILL
37
21028

SPORTS
SOUP'S
GRILL
37

SOUTH ST.
BURGER CO.

ST. BURGER CO.
"The best onion rings I
- Barrie Examiner 100% BEEF
are first rate...far superior to the drive
FIRE GRILLED
"Habit-forming milkshakes" - Toronto Life
NO HEAT LAMPS
TOTALLY OBSESSED cooked in sunflower oil
hand made ONION RINGS

σουβlike
ΛΑΔΙ
ΠΙΠΕΡΙ
ΣΚΟΡΔΟ
WC
Coca-Cola
Coca-Cola
Coca-Cola

σουβlike

σουβlike
ΡΙΓΑΝΗ
ΦΕΤΑ
ΠΑΤΑΤΕΣ
ΛΑΔΙ
ΠΙΠΕΡΙ
ΣΚΟΡΔΟ
ΛΑΚΩΝΙΑ
ΑΡΙΔΑΙΑ
ΤΡΙΠΟΛΗ
VIRGIN
OLIVE OIL
NISSA

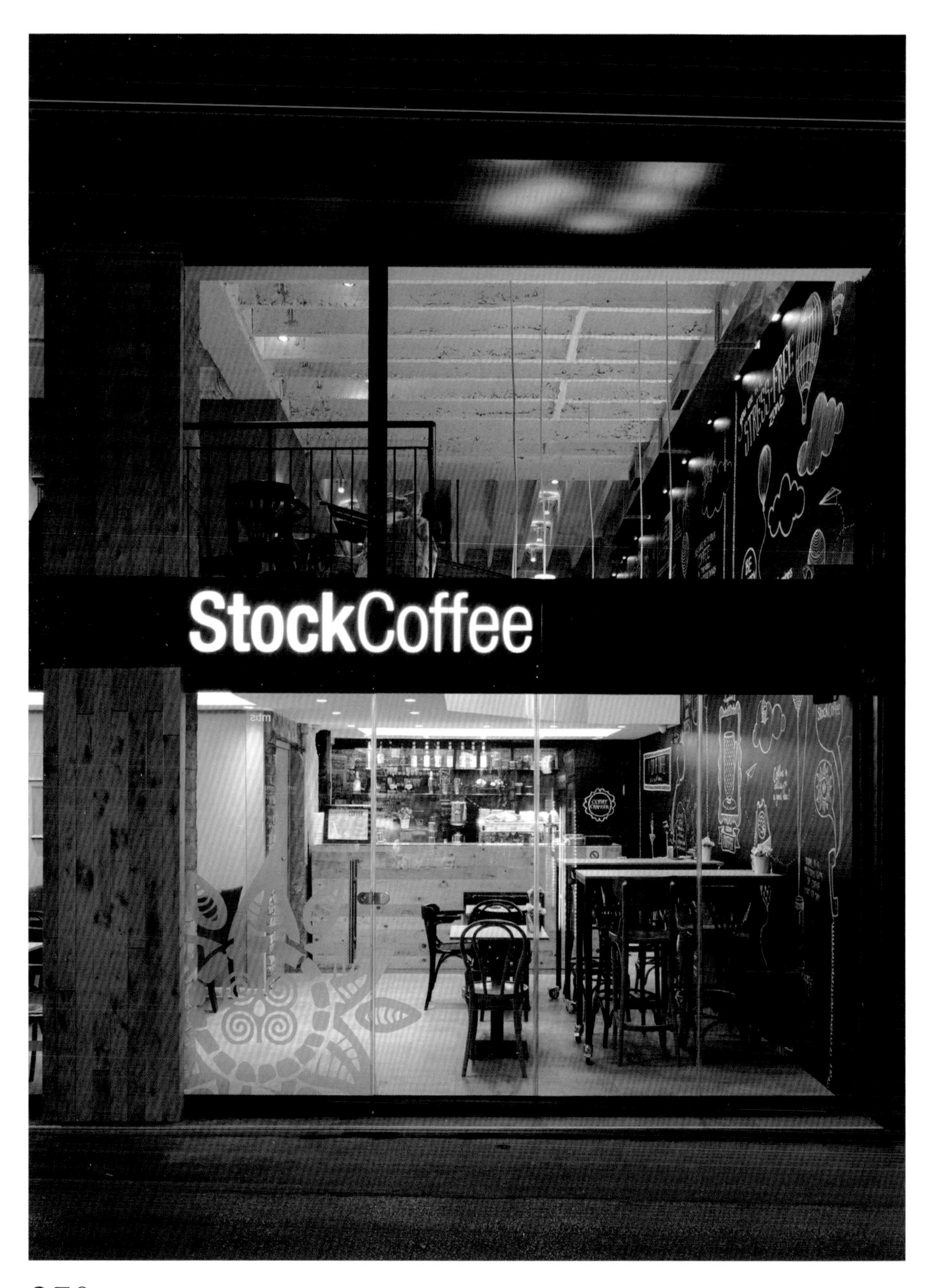
StockCoffee

StockCoffee

the
doner
company
the
doner
company

TOMASELLI

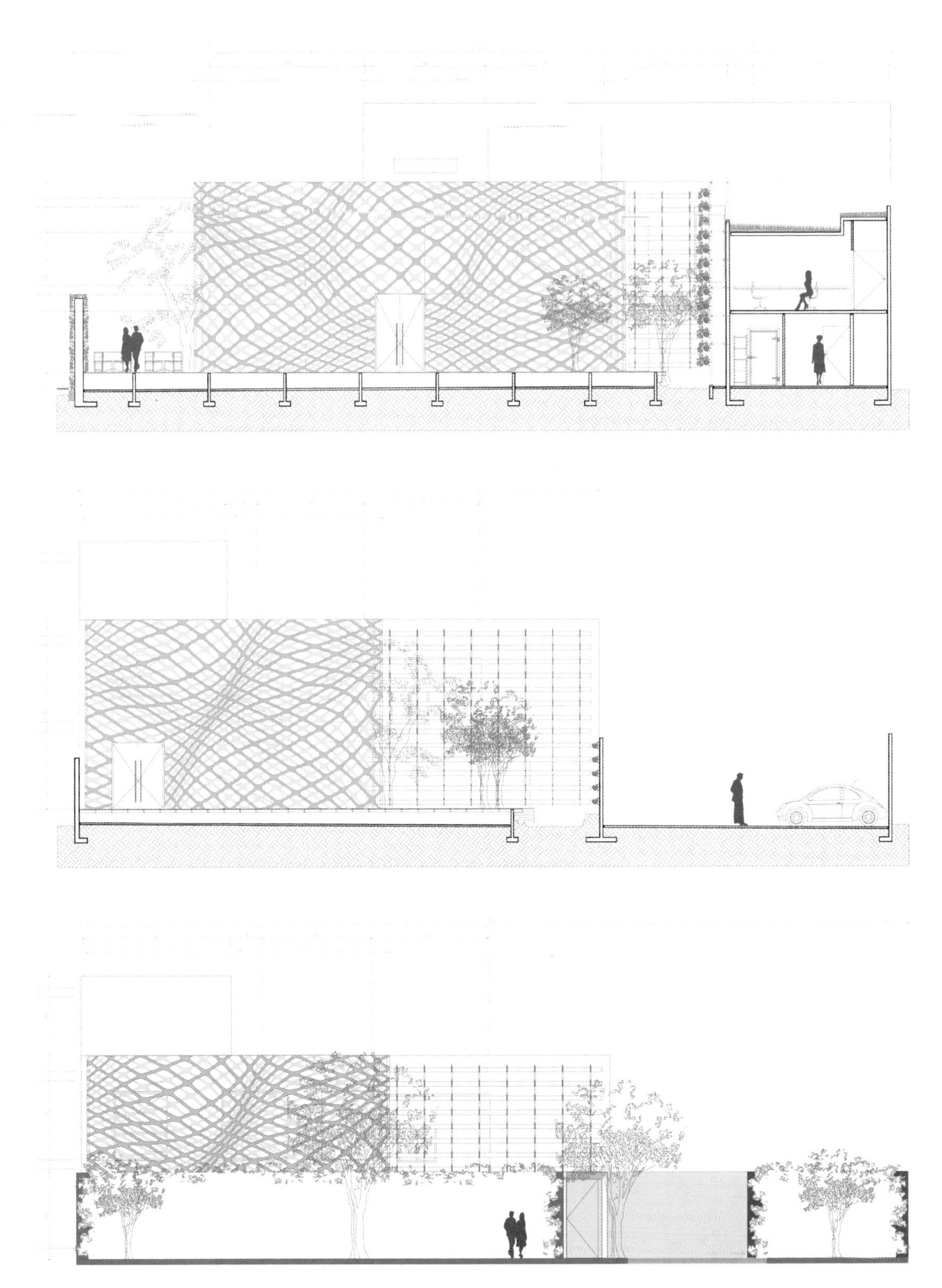

TSUJITA LA
ARTISAN NOODLES

TSUJITA LA
ARTISAN NOODLES

TSUJITA LA
ARTISAN NOODLES

TSUJITA LA

TSUJITA LA
ARTISAN NOODLES

UNITED
CHICKEN

UNITED
CHICKEN

UNITED
CHICKEN
UNITED
KIDS

UNITED
CHICKEN

vesu

waku
waku

waku
waku

176
水
WATERMOON
月

燒肉达人
Yakinjku Master Restaurant

Index
索引

Index
索引

图书在版编目（CIP）数据

餐厅店面 / 新空间编辑组编 ; 张晨译.
- 沈阳 : 辽宁科学技术出版社, 2016.3
ISBN 978-7-5381-9553-8

Ⅰ. ①餐… Ⅱ. ①新… ②张… Ⅲ. ①餐厅－室内装饰设计－世界－图集 Ⅳ. ①TU247.3-64

中国版本图书馆 CIP 数据核字(2016)第 013614 号

出版发行：辽宁科学技术出版社
（地址：沈阳市和平区十一纬路29号 邮编：110003）
印 刷 者：利丰雅高印刷（深圳）有限公司
经 销 者：各地新华书店
幅面尺寸：170mm×225mm
印　　张：19
插　　页：4
出版时间：2016年 3 月第 1 版
印刷时间：2016年 3 月第 1 次印刷
责任编辑：殷　倩
封面设计：周　洁
版式设计：周　洁
责任校对：周　文

书　　号：ISBN 978-7-5381-9553-8
定　　价：88.00元

联系电话：024-23284360
邮购热线：024-23284502
http://www.lnkj.com.cn